PONIES

Assateague Island's Mane Attraction

Salty Snack. The wild ponies feed on a salty plant called cordgrass. Every spring, foals are born on Assateague Island. The baby ponies graze on cordgrass too.

Soon cowboys herd the ponies through town to the carnival grounds. The next day most of the young ponies, called **foals**, will be auctioned, or sold to the highest bidder. The pony auction does three things:

- It raises money for the Chincoteague Fire Department.
- It allows some people to take home a foal.
- And most important, it keeps the pony population at the proper size. Resources such as food will support only about 150 ponies on the southern end of Assateague Island. A larger number would hurt the island's **ecology**, or balance of life.

History and Mystery

Assateague is a long, narrow island. It stretches between southern Maryland and northern Virginia. On one side is the Atlantic Ocean. On the other side is a quiet bay.

The ponies have been roaming free on the island for hundreds of years. They are **feral** animals. This means that their ancestors once were tame.

No one knows exactly how the ponies got to the island. Some people believe that long ago the first ponies were transported by ship from Spain. They think the ship wrecked near the island in a storm, and the ponies swam ashore.

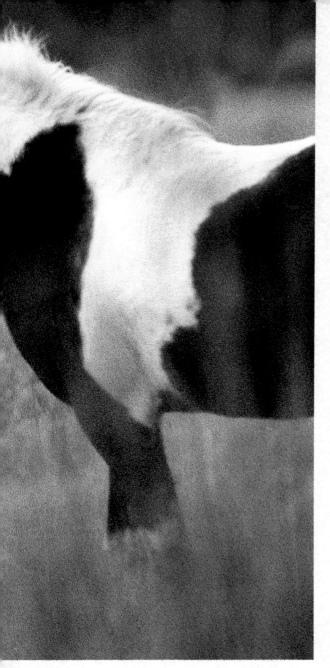

Most experts, though, think the first settlers of mainland Maryland and Virginia brought the ponies with them from England. Later they turned the animals loose to graze on Assateague Island.

Harsh Habitat

Today's ponies lead a hard life. In the summer they face hot weather and biting insects. In the winter they must grow thick coats to protect themselves from bitter winds.

Spring and fall are the best seasons. The weather on the island is mild, and there is plenty of grass for the ponies to eat. The ponies also eat leaves and twigs. They even munch on poison ivy, which doesn't seem to bother them.

These island grazers may be the size of ponies—less than 147 centimeters (58 inches) tall—but they are actually horses. Experts think that the harsh **habitat**, or place where they live, accounts for their small size. In fact, when some of the auctioned foals leave Assateague and receive better food and shelter, they grow to horse size. But people have been calling them ponies for years, and the name has stuck.

Pony Bands

The ponies live together in small groups called **bands**. Some bands may have as few as two ponies. Others may have a dozen. In most bands there are usually several **mares**, or adult females, some foals, and one adult male.

The adult male pony is called a **stallion**. It is his job to protect the band. Sometimes stallions try to steal ponies from other bands. This can lead to fights between stallions. They bite and kick with their heavy hooves until one stallion backs away.

In the spring mares give birth. Within minutes, their foals begin to walk on wobbly legs. Soon they are running and playing. At first they drink their mother's milk to help them grow. Then they begin to eat grass as the older ponies do.

Managing the Herds

There are two main groups, or **herds**, of wild ponies on Assateague Island. Each herd has 100 to 150 ponies and includes many pony bands. One herd lives on the Maryland side of the island. The other lives on the Virginia side. A fence at the state line keeps the herds apart.

National Park Service rangers manage the herd on the Maryland side of the island. They control the number of ponies there by using a special **vaccine**, or medicine. Each year they inject the vaccine into some of the mares. The vaccine keeps the mares from having babies that year.

The Chincoteague Fire Department manages the herd on the Virginia side of the island. It controls the number of ponies there through the annual pony auction.

Return to the Wild

At the auction, some people bid on ponies to take home. Others just come to watch. In 2001, 85 ponies were sold. One foal sold for $10,500. That set a new record price for the auction. In all, that auction raised $167,000.

The day after the auction, Chincoteague cowboys herd the ponies back to the water's edge. Crowds cheer again as the ponies swim home to Assateague Island. There they will be free to roam again for another year.

On the Fence. It's not always easy to choose a favorite foal at the auction.

Wordwise

band: small group of horses

channel: waterway between two landmasses that lie close to each other

ecology: how plants and animals live in relation to each other

feral: wild animals whose ancestors were once tame

foal: young horse

habitat: place where something lives

herd: large group of horses (often contains many bands)

mare: adult female horse

pony: small horse that is less than 147 centimeters (58 inches) tall when fully grown

stallion: adult male horse

vaccine: medicine swallowed or injected into the body

Pony Parts

The Assateague ponies may be small compared to other horses. Yet they have had big success in their island home. These tough ponies have features, or parts, that help them survive in the wild.

Eyes
A pony's eyes are on the sides of its head. When a pony is standing still, the only places it can't see are the spots directly in front and behind it.

Tail
Ponies can swing their tails to swat flies and other pesky insects off their bodies.

Hair
In summer, the ponies have thin coats of short, silky hair. In winter, their coats get thick and keep them warm in cold weather.

Legs
Ponies have long legs that let them run quickly away from danger. Long legs are also good for walking through tall grass and bushes.

Hooves
Each hoof has a hard covering. This lets ponies walk or run safely on many surfaces.

Island Refuges

Piping plover

Fly By. Many of the islands' feathered friends don't live on the islands. They visit Chincoteague and Assateague when they are flying to their seasonal homes.

Chincoteague and Assateague are famous for their ponies. Yet this pair of islands is home to an amazing variety of other animals. Why do so many creatures thrive on these small islands? Certain sections of the island are protected. That means there are rules about what people can do on the land—and what they can't do.

Safe Spaces

The Chincoteague National Wildlife Refuge is one of these protected patches. It includes more than 5,666 hectares (14,000 acres) of land on both Chincoteague and Assateague Islands.

The refuge provides a safe space for plants and animals. It protects their habitats, or the places where they live. The refuge offers plenty of natural shelters too. Animals find homes among the pine trees, in the soggy marshes, and within dry sand dunes.

The islands make nice year-round homes for animals that hunt in the ocean. Many birds nest on the islands. During the day, they catch fish in the ocean. At night, the birds return to the islands to sleep.

Flocking to the Islands

The islands are also popular vacation spots for animals on the move. Every year, flocks of birds migrate, or travel long distances between their winter and summer homes.

These birds live in the north during the summer. When the weather turns cold, they travel south. They fly thousands of miles to warm winter homes. In spring, they fly back along the same path.

So many birds travel along the eastern coast of North America that scientists call it the Atlantic Flyway. It's like a highway for birds. But flying for days on end can wear the birds out. Where do they stop to rest? You guessed it—Chincoteague and Assateague Islands.

Sika deer

Willet

Squirrel

Wet and Wild

Life can get pretty wild on the islands. Chincoteague and Assateague are barrier islands. They provide a barrier that protects the mainland from rough ocean waves. Wind whips the sand from place to place. Tides sweep the sand away. Yet even in this harsh habitat, many living things find a way to thrive. Tough grasses grow in the dunes. Ghost crabs burrow in the sand.

Saltwater marshes are another kind of habitat on the islands. These are areas where ocean water covers the land. Many kinds of shellfish and other small animals live in this salty habitat. Black ducks and other migratory birds depend on these critters for food.

Life in the Forest

Chincoteague and Assateague are flat and windswept. Yet on the highest ground, you can also find forests. Here, the land doesn't flood with salty ocean water as it does on the beach or in the marshes. So these areas have a very different look. Tall pine trees tower above the sandy soil.

Instead of water-loving birds and side-stepping crabs, the forests are home to other animals. Squirrels leap from tree to tree. Deer and wild ponies nibble on the bushes and plants. Foxes and raccoons roam the forest floor in search of food. Owls nest in the treetops. At night they swoop down to scoop up mice and other tasty treats.

New Horizons. The island refuge is a protected area. It gives new life to animals and plants that are struggling to survive.

Saving Species

The Chincoteague Island National Wildlife Refuge does more than just protect the land for animals. It also helps species survive. For example, bald eagles and piping plovers are struggling to survive in the wild. They get a boost by living on these special islands. The refuge gives them a safe habitat. Here the birds can find food and a safe place to raise their young.

Ponies may be the superstars of these two islands in the Atlantic Ocean. But it's the variety of wildlife that makes the islands so amazing. Many kinds of animals live on these islands. Here ponies run wild and the animals outnumber the people.

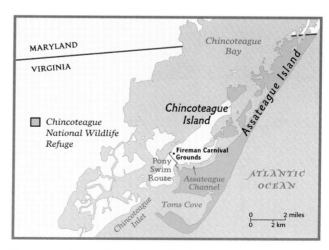

Protecting the Land. Chincoteague and Assateague are barrier islands. They protect the mainland from rough ocean waves.

ISLAND LIFE

Take a run at these questions to
see how much you've learned.

1 Why do ponies swim from
Assateague Island to
Chincoteague Island?

2 Are the ponies on Assateague
Island really ponies? Explain.

3 Is life easy for the Assateague
ponies? Why or why not?

4 Why do people control the
number of ponies on the islands?

5 Why do so many kinds of animals
come to the islands?